MACK EH-EJ-EM-EQ-ER-ES
1936 THROUGH 1950
PHOTO ARCHIVE

MACK EH-EJ-EM-EQ-ER-ES 1936 THROUGH 1950

PHOTO ARCHIVE

Photographs from the
Mack Trucks Historical Museum Archives

Edited with introduction by
Thomas E. Warth

Iconografix
Photo Archive Series

Iconografix
PO Box 609
Osceola, Wisconsin 54020 USA

Library of Congress Card Number 95-77487

ISBN 1-882256-39-5

95 96 97 98 99 00 5 4 3 2 1

Cover and book design by Lou Gordon, Osceola, Wisconsin

Printed in the United States of America

Book trade distribution by Voyageur Press, Inc. (800) 888-9653

PREFACE

The histories of machines and mechanical gadgets are contained in the books, journals, correspondence and personal papers stored in libraries and archives throughout the world. Written in tens of languages, covering thousands of subjects, the stories are recorded in millions of words.

Words are powerful. Yet, the impact of a single image, a photograph or an illustration, often relates more than dozens of pages of text. Fortunately, many of the libraries and archives that house the words also preserve the images.

In the *Photo Archive Series*, Iconografix reproduces photographs and illustrations selected from public and private collections. The images are chosen to tell a story—to capture the character of their subject. Reproduced as found, they are accompanied by the captions made available by the archive.

The Iconografix *Photo Archive Series* is dedicated to young and old alike, the enthusiast, the collector and anyone who, like us, is fascinated by "things" mechanical.

ACKNOWLEDGMENTS

The photographs appearing in this book were made available by the Mack Trucks Historical Museum. We are grateful to Colin Chisholm, Curator, for his assistance.

Mack EH. April 1936. (A7072)

INTRODUCTION

"Built like a Mack Truck." What a wonderful statement—one of the best descriptions you can give to a piece of equipment designed to stand up to tough conditions. Since just after the turn of the century, Mack has been turning out trucks of such a quality that the phrase has become part of our language. The first truck placed in the Smithsonian Collection was a Model AC—the venerable "Bulldog."

The Mack brothers made their name as horse drawn wagon builders in Brooklyn, New York in the 1800s. About 1902 they produced their first motor vehicle, and by 1911 Mack produced over 500 trucks a year. The outbreak of World War I proved a boon to their business, which by then was established in its present headquarters of Allentown, Pennsylvania.

By the mid-1930s, the United States' economy had begun its recovery from the Great Depression. The role of truck transportation continued to evolve, and both engineering and styling changes were apparent in Mack models introduced in the late 30s. Following the lead of the automobile industry, smaller trucks became more streamlined.

MACK EH, EJ, EM, EQ, ER, ES 1936-1950 Photo Archive features the medium-duty models in the E range. The photographs were chosen from the Mack Trucks Historical Museum and are presented in chronological order within models. Captions are those found on the illustrations and negative numbers are given when known. "1D" indicates one drive axle; "2D" indicates two drive axles; "SW" is six wheels; "T" indicates tractor unit; "U" indicates cab-over-engine; "I" is single axle, and "S" single reduction.

We hope that by presenting these fascinating images the reader will be encouraged into further research.

MODEL EH

In terms of numbers of units sold and prior to the B-Series of the fifties and sixties, this model was Mack's most successful after the AB and Bulldog. 31,539 were built between 1936 and 1950. Gross Vehicle Weight (GVW) ranged from 18,000 and 19,500 pounds on the earliest versions, to 40,000 pounds combined tractor and trailer weight on the later models. Engines ranged in size from the 310-cubic inch gasoline to the 457-cubic inch diesel.

EH chassis. April 1936. (A7071)

EH tractor. June 1936. (M1658)

EH tractor. August 1936. (M1719)

EH six-wheeler, showing transfer case and twin secondary driveshaft. July 1936. (A7241)

14

EH six-wheeler. July 1936. (A7244)

EH tractor and GE traveling display trailer. October 1936. (M1794)

EH trucks with prison van bodies. October 1936. (M1809)

EH. October 1936. (M1819)

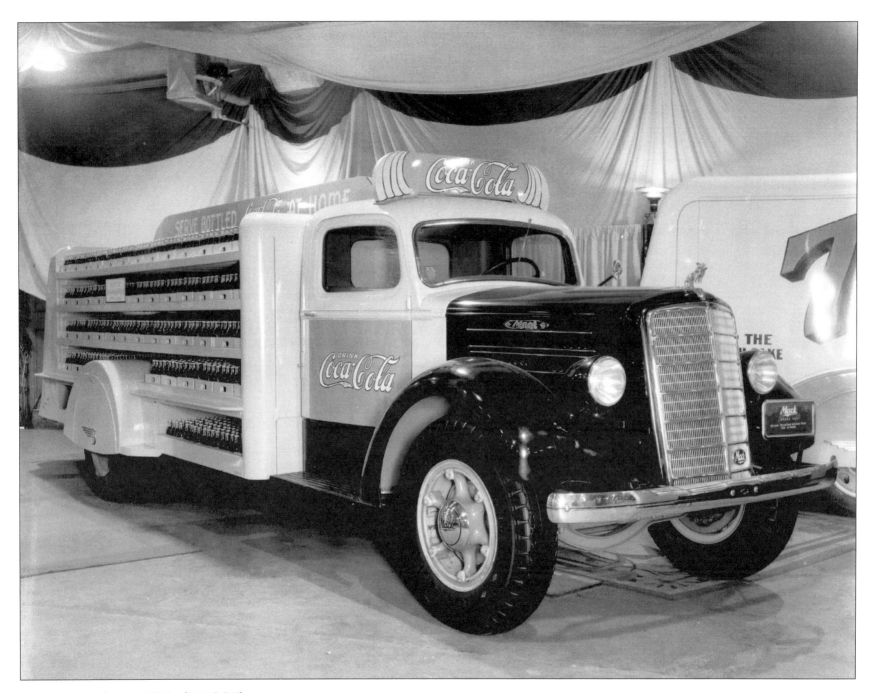

EH. November 1936. (M1907)

EH tractor. November 1936. (M1919)

EH. December 1936. (M1987)

EH. January 1937. (M1999)

22

EH tractor. February 1937. (A7637)

EH tractor. March 1937. (M2178)

EH. April 1937. (M2190)

EH diesel six-wheeler, chassis #2338, delivered July 1937. Tyre Bros. Glass, Los Angeles, California. (1159)

EH diesel tractor, chassis #1314, delivered October 1937. (5237)

EH. October 1937. (A8163)

EHU cab-over-engine. December 1937. (M2710)

EHU cab-over-engine. April 1938. (A8591)

EH. June 1938. (M3103)

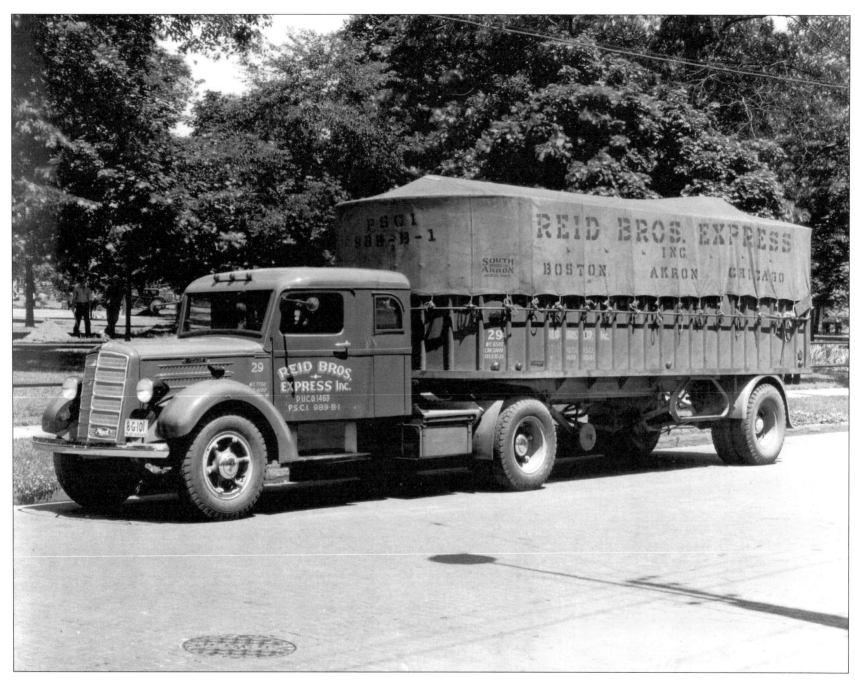

EH tractor. June 1938. (M3135)

EH with snow plow. August 1938. (M3248)

EH tractor. August 1938. (M3273)

EHU cab-over-engine. September 1938. (A8881)

Ten EH six-wheelers and three EJs work on the Marine Corps airport at Quantico, Virginia. December 1938.
(M3701)

EH. March 1939. (A9230)

EHU cab-over-engine. April 1939. (M3963)

EH tractor. May 1939. (A9381)

EH. August 1939. (M4250)

EHU cab-over-engine tractor. October 1939. (M4408)

EH. November 1939. (M4492)

EHU cab-over-engine. November 1939. (M4511)

EHSW six-wheeler. January 1940. (M4537)

EH tractor. May 1940. (M4799)

EH tractors. June 1940. (M4854)

EHT tractor. July 1940. (M9041)

AB, early series B, EH tractor, EHU cab-over-engine tractor, and Packard. July 1940. (M4956)

EHUT cab-over-engine tractor. November 1940. (M5317)

EHUT cab-over-engine tractor, chassis #1324. August 1941. (V3232)

EHUT cab-over-engine tractor. August 1941. (M5933)

EH. December 1941. (M6256)

EHU cab-over-engine used to transport workers in Iran. May 1942. (M6504)

EH1D, chassis #6804. May 1944. (M7896)

EHT tractor, Row River Lumber Co., Dorena, Oregon. (M8755)

EHT and LJT tractors, Alabama Highway Express Inc. May 1945. (M8849)

EHT tractor transporting 42-ton tug for use in Gulf Coast oil fields. July 1945. (M9026)

EH tractor in Birmingham, Alabama War Bond Parade. August 1945. (M9066)

EHIS, Consumers Steel & Supply, Racine, Wisconsin. March 1946. (M9442)

EHX1D dump truck, chassis #8969. March 1946. (V9207)

EH. April 1946. (V9323)

EH1D, chassis #8877, Eichler's Brewery, Bronx, New York. May 1946. (M9554)

62

EH. May 1946. (M9563R)

EHT tractor. Magnolia Petroleum Co., Houston, Texas. May 1946. (M9583)

EHT tractors. June 1946. (M9630)

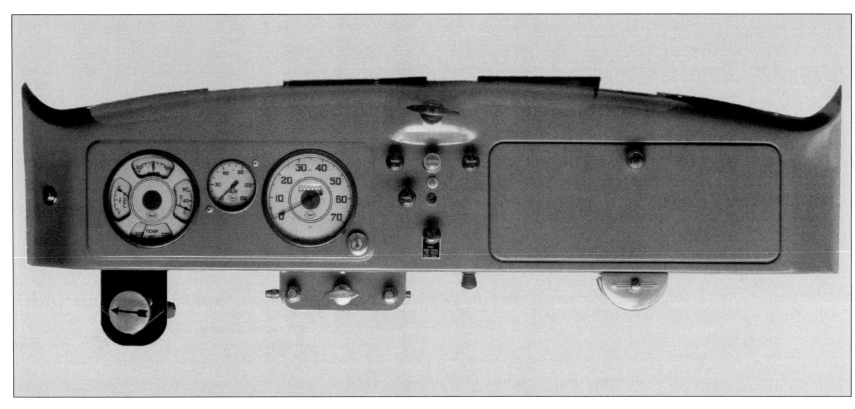

EH composite instrument panel. July 1946. (V9332)

EH. July 1946. (V9341)

EH converted into tractor, Brown & Root, Houston, Texas. June 1946. (M9660)

EHU and LMU cab-over-engines. July 1946. (M9676)

EHT tractor. July 1946. (M9709)

EHT tractor. July 1946. (M9711)

EHU cab-over-engines. July 1946.

EH. Kansas City Public Service Company street car and bus wrecker. September 1946. (M9779)

EHT tractor. January 1947. (V9573)

EHX1D dump truck, chassis #9050. April 1947. (V9907)

EHT tractors. August 1947. (M11459)

EHU cab-over-engine fleet. September 1947. (M11523)

EHU cab-over-engine, George M. Brewster & Son Inc., Bogata, New Jersey. June 1948. (M12272)

EHX dump truck. June 1948. (M12219)

EHT1D tractor, chassis #6316. October 1948. (C3169)

EHU cab-over-engine. October 1948. (M12610)

EH. December 1948. (M12691)

EHUIS cab-over-engine, chassis #1859. May 1949. (C3774)

EHT1D tractor, chassis #10638.
October 1949. (M14196)

EHU cab-over-engine tractor.

EH dump truck with snow plow.

EHU cab-over-engine fleet. (715)

MODEL EJ

762 built in 1937 and 1938. GVW was 16,000 pounds with a 288-cubic inch gasoline engine.

EJ. June 1937. (A7869)

MODEL EM

1,584 built between 1937 and 1943. GVW was 23,000 pounds with a 310-cubic inch gasoline engine and a 354-cubic inch gasoline engine in the tractor version.

EMU cab-over-engine dump truck. August 1939. (M4309)

EMUT1D cab-over-engine tractor, chassis #1146. May 1942. (V4201)

EM. May 1939. (A9424)

MODEL EQ

10,661 were built between 1937 and 1950. Original cab-over-engine (EQU) GVW was 16,000 pounds with the EQ at 23,000 pounds. By the late forties, design improvements increased GVW to 25,000 pounds on the dump truck, 27,000 pounds on the EQ, and 45,000 pounds on the tractor-trailer combination. The 354-cubic inch gasoline engine was used in the original specifications, and was superseded by the 431-cubic inch Thermodyne engine. The 457-cubic inch and 510-cubic inch diesel engines were optional.

EQ. April 1937. (A7748)

First production EQ. February 1937. (A7704)

EQ. June 1937. (A7889)

EQ tractor. October 1937. (M2587)

EQ. February 1938. (M2867)

EQ. August 1938. (M3288)

EQSW six-wheeler, chassis #1245, Lockett & Son., Pasadena, California. April 1938.

EQ tractor. March 1939. (A9264)

EQ tractor. May 1939. (M4036)

EQ tractor. August 1939. (M4353)

104

EQ tractor. October 1939. (M4428)

EQU cab-over-engine fleet. February 1940. (V1197)

EQU cab-over-engine dump truck. February 1940. (V1199)

EQ. February 1940. (M5499)

EQT tractor. October 1941. (M6070)

EQX dump truck in a coal stripping operation. Hazleton, Pennsylvania. July 1945. (V8481)

EQX1D, chassis #5000. April 1947. (C1031)

"*Mack*, the purebred English bull dog, is taking it all very seriously in this affectionate pose with pretty Miss Holly Hamill. *Mack* was won by Mr. Gordon McCaig, Aviation Operations Manager, Shell Oil Co. of Canada Ltd., holder of the lucky number in the contest conducted by Mack Trucks of Canada Ltd. on the occasion of the Automotive Transport Association of Ontario's Annual Dinner." (Posed with a Model EQ)

EQX dump truck fleet, Tellyer Concrete Pipe Co., New Mexico. July 1947. (M11343)

EQ2D six-wheeler. December 1947. (C2024)

EQSW six-wheeler, Continental Oil Co., Salt Lake City, Utah. April 1948. (M12062)

EQSW six-wheeler. April 1948. (M12067)

EQSW six-wheeler. July 1948. (M12328)

EQU2D six-wheeler, chassis #1185. September 1948. (C3032)

EQT tractor pilot model. April 1949. (C3695)

EQSW six-wheeler. September 1949. (C4123)

EQSW six-wheeler. September 1949. (C4133)

MODEL ER

At 20,000 pounds GVW, this version was designed as a replacement for the venerable AB. 359 were built between 1936 and 1941, mostly equipped with the 310-cubic inch gasoline engine.

ER dump truck, chassis #1007, A. E. Fowler & Sons, Santa Ana, California. January 1937. (1129)

MODEL ES

This version was basically a chain-driven version of the EQ. 75
were built between 1938 and 1940.

ES. March 1940. (V1283)

BIBLIOGRAPHY

Brownell, Tom, *History of Mack Trucks*, Osceola, Motorbooks International, 1994.

Montville, John B., *Mack*, Newark, Walter Haessner, Inc., 1973.

Montville, John B., *Mack, A Living Legend of the Highway*, Tucson, Aztex Corp., 1979.

Rasmussen, Henry, *Mack, Bulldog of American Highways*, Osceola, Motorbooks International, 1987.

The Iconografix Photo Archive Series includes:

TRACTORS AND CONSTRUCTION EQUIPMENT

CASE TRACTORS 1912-1959 Photo Archive	ISBN 1-882256-32-8
CATERPILLAR MILITARY TRACTORS VOLUME 1 Photo Archive	ISBN 1-882256-16-6
CATERPILLAR MILITARY TRACTORS VOLUME 2 Photo Archive	ISBN 1-882256-17-4
CATERPILLAR SIXTY Photo Archive	ISBN 1-882256-05-0
CATERPILLAR THIRTY Photo Archive	ISBN 1-882256-04-2
FARMALL F-SERIES Photo Archive	ISBN 1-882256-02-6
FARMALL MODEL H Photo Archive	ISBN 1-882256-03-4
FARMALL MODEL M Photo Archive	ISBN 1-882256-15-8
FARMALL REGULAR Photo Archive	ISBN 1-882256-14-X
FORDSON 1917-1928 Photo Archive	ISBN 1-882256-33-6
HART-PARR Photo Archive	ISBN 1-882256-08-5
HOLT TRACTORS Photo Archive	ISBN 1-882256-10-7
JOHN DEERE MODEL A Photo Archive	ISBN 1-882256-12-3
JOHN DEERE MODEL B Photo Archive	ISBN 1-882256-01-8
JOHN DEERE MODEL D Photo Archive	ISBN 1-882256-00-X
JOHN DEERE 30 SERIES Photo Archive	ISBN 1-882256-13-1
MINNEAPOLIS-MOLINE U-SERIES Photo Archive	ISBN 1-882256-07-7
OLIVER TRACTORS Photo Archive	ISBN 1-882256-09-3
RUSSELL GRADERS Photo Archive	ISBN 1-882256-11-5
TWIN CITY TRACTOR Photo Archive	ISBN 1-882256-06-9

TRUCKS

DODGE TRUCKS 1929-1947 Photo Archive	ISBN 1-882256-36-0
DODGE TRUCKS 1948-1960 Photo Archive	ISBN 1-882256-37-9
MACK MODEL AB Photo Archive	ISBN 1-882256-18-2
MACK MODEL B 1953-66 Photo Archive	ISBN 1-882256-19-0
MACK MODEL B 1953-1966 VOLUME 2 Photo Archive	ISBN 1-882256-34-4
MACK EB, EC, ED, EE, EF, EG & DE 1936-1951 Photo Archive	ISBN 1-882256-29-8
MACK EH-EJ-EM-EQ-ER-ES 1936-1950 Photo Archive	ISBN 1-882256-39-5

MACK FC, FCSW & NW1936-1947 Photo Archive	ISBN 1-882256-28-X
MACK FG-FH-FJ-FK-FN-FP-FT-FW 1937-1950 Photo Archive	ISBN 1-882256-35-2
MACK LF-LH-LJ-LM-LT 1940-1956 Photo Archive	ISBN 1-882256-38-7
STUDEBAKER TRUCKS 1927-1940 Photo Archive	ISBN 1-882256-40-9
STUDEBAKER TRUCKS 1941-1964 Photo Archive	ISBN 1-882256-41-7

AUTOMOTIVE

AMERICAN SERVICE STATIONS 1935-1943 Photo Archive	ISBN 1-882256-27-1
IMPERIAL 1955-1963 Photo Archive	ISBN 1-882256-22-0
IMPERIAL 1964-1968 Photo Archive	ISBN 1-882256-23-9
LE MANS 1950: THE BRIGGS CUNNINGHAM CAMPAIGN Photo Archive	ISBN 1-882256-21-2
SEBRING 12-HOUR RACE 1970 Photo Archive	ISBN 1-882256-20-4
STUDEBAKER 1933-1942 Photo Archive	ISBN 1-882256-24-7
STUDEBAKER 1946-1958 Photo Archive	ISBN 1-882256-25-5

AVAILABLE EARLY 1996

CLETRAC AND OLIVER CRAWLERS Photo Archive	ISBN 1-882256-43-3
COCA-COLA: A HISTORY IN PHOTOGRAPHS 1930-1969	ISBN 1-882256-46-8
COCA-COLA: ITS VEHICLES IN PHOTOGRAPHS 1930-1969	ISBN 1-882256-00-X
FARMALL SUPER SERIES Photo Archive	ISBN 1-882256-49-2
INTERNATIONAL TRACTRACTORS Photo Archive	ISBN 1-882256-48-4
PACKARD 1935-1941 Photo Archive	ISBN 1-882256-44-1
PACKARD 1942-1958 Photo Archive	ISBN 1-882256-45-X
PHILLIPS 66 1945-1953 Photo Archive	ISBN 1-882256-42-5
SEBRING 12 HOUR RACE 1962 Photo Archive	ISBN 1-882256-51-4
SEBRING 12 HOUR RACE 1965 Photo Archive	ISBN 1-882256-50-6

The Iconografix Photo Archive Series is available from direct mail specialty book dealers and bookstores worldwide, or can be ordered from the publisher. For additional information or to add your name to our mailing list contact:

Iconografix
PO Box 609
Osceola, Wisconsin 54020 USA

Telephone: (715) 294-2792
(800) 289-3504 (USA and Canada)
Fax: (715) 294-3414

Book trade distribution by Voyageur Press, Inc. (800) 888-9653

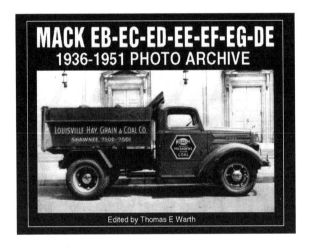

MACK MODEL AB
PHOTO ARCHIVE

MILL'S DRAYAGE Cº

Edited by Thomas E. Warth

MORE
GREAT BOOKS ABOUT
MACK TRUCKS

MACK MODEL AB *Photo Archive*
ISBN 1-882256-18-2
MACK MODEL B 1953-1966 VOLUME 1
Photo Archive ISBN 1-882256-19-0
MACK EB, EC, ED, EE, EF, EG & DE
1936-1951 *Photo Archive*
ISBN 1-882256-29-8
MACK FC, FCSW & NW1936-1947
Photo Archive ISBN 1-882256-28-X
MACK MODEL B 1953-1966
VOLUME 2 *Photo Archive*
ISBN 1-882256-34-4
MACK EH-EJ-EM-EQ-ER-ES 1936-1950
Photo Archive ISBN 1-882256-39-5
MACK LF-LH-LJ-LM-LT *1940-1956*
Photo Archive ISBN 1-882256-38-7

**Available from your favorite
bookseller or from Iconografix.
To order by phone:
(800) 289-3504 (US and Canada)
or (715) 294-2792.**

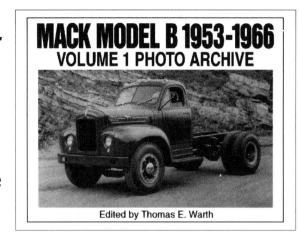

MACK MODEL B 1953-1966
VOLUME 1 PHOTO ARCHIVE

Edited by Thomas E. Warth

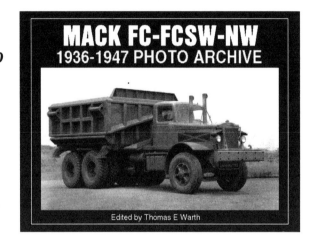

MACK FC-FCSW-NW
1936-1947 PHOTO ARCHIVE

Edited by Thomas E Warth

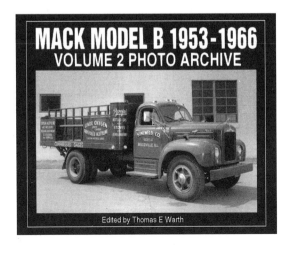

MACK MODEL B 1953-1966
VOLUME 2 PHOTO ARCHIVE

Edited by Thomas E Warth

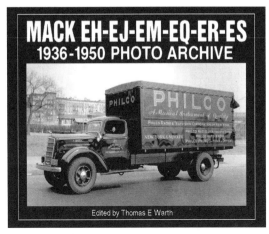

MACK EH-EJ-EM-EQ-ER-ES
1936-1950 PHOTO ARCHIVE

PHILCO

Edited by Thomas E Warth

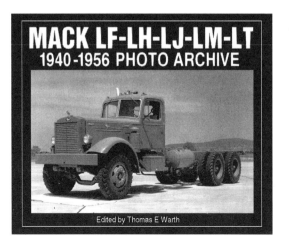

MACK LF-LH-LJ-LM-LT
1940-1956 PHOTO ARCHIVE

Edited by Thomas E Warth